Mixing cat wisdom with science and solar politics

THE RETURN OF THE
SOLAR CAT
BOOK

JIM AUGUSTYN

with illustrations by Hildy Paige Burns

Patty Paw Press • Berkeley, California

Library of Congress Control Number: 2003091511
ISBN: 0-9729949-0-4

Book and cover designed by Kathleen Thorne Thomsen
Printed in Singapore

PATTY PAW PRESS
P.O. Box 7075
Berkeley, California 94707
U.S.A.
www.solarcat.com

FOR LITTLE MISS PATTY PAW

Original Thanks

To Priscilla Thomas, Jack Howell. Bill Malenius, Steve Burns, Alana Shindler, Carol David, Margaurite Segal, Tony Wexler, Judy Jennings, Bruce Wilcox, Chip Barnaby, Max Jacobsen, Murray Silverstein, Michael Floyd, Nancy Erlich, Juanita Rusev, Sylvia Grissom, Penny Niland, and Marshal Hunt.

Return Thanks

Robin Mitchell, Rob Nelson, Martha McEvoy, Taylor Geer, Tom Stoffel, Chuck Kutschner, Ken Haggard, Polly Cooper, Chip Tittmann, Sarah Tittmann, Carin Gummo, Dave Kearney, Paul Nava, Bob Cable, Mary Jane Hale, JoAnn Fitch, (the real) John Schaefer, Glen Friedman, Don Osborne, Mike Nicklas, Denis Hayes, Dan Shugar, Tom Dinwoodie, Charlie Huizenga, John Dunlop, Karen Conover, John Gould, and Tom Burton.

Extra Special Thanks

To Liz, Katie, Kathleen, and Hildy. And to Mouse, Parsley, Emily, Jenny, Max, Ernie, their ancestors, and their successors.

About the Author

Jim Augustyn was born and lived in Chicago until graduating from the Illinois Institute of Technology with a B.S. degree in mechanical and aerospace engineering. He moved to San Francisco and then Berkeley where he now resides with his wife Elizabeth and their two cats: Max and Ernie.

His engineering career has taken him from nuclear power plant design to computer simulation of energy use in buildings to solar energy research and development for government and industry.

His literary career has focused entirely on solar cats.

TABLE OF CONTENTS

Preface to the Return Edition   vii

1. Cats and the Sun   1

2. Solar Thermal Cats   16

3. Solar Electric Cats   39

4. Solar Cat Economics   58

5. Choosing a Solar Cat   68

6. Sustaining a Solar Cat   73

Appendices:

Appendix 1. *Active Solar Thermal Cat Economic Calculations*   78

Appendix 2. *"SOLCAT" Solar Cat Evaluation Computer Program*   80

Appendix 3. *Cat Thermal Units (CTUs) and Heat*   82

Appendix 4. *Concrete Cats—The Allergic Solution*   84

Appendix 5. *Glossary*   86

Appendix 6. *Recommended Reading Material*   88

Though people still have much to learn about solar energy, with cats to guide them, they are assured of eventual success.

PREFACE TO THE RETURN EDITION

• Outdoor cats who come and go as they please at times disappear from their homes for long periods. Concerned about their safety, their human friends frantically search their neighborhoods, calling their cat's names as they post "lost kitty" signs. While such fears may be understandable, they are needless, for cats *always* return. • *The Solar Cat Book* first appeared in 1979, and although it never really disappeared, it has now returned. • Those who memorized that first edition of *The Solar Cat Book* will find the *Return Edition* contains most of the original material. It also contains new material reflecting actual human progress in a few areas of solar energy. This means you need only memorize the different parts to stay up-to-date. • And to those who ask if reading this book will do them any good, first ask what *is* good? If you can answer that one, it should be a simple matter to know if you actually have anything better to do. •

Jim Augustyn
Berkeley, California
2003.02.14

The idea of a solar cat has a firm basis in reality familiar to all who know cats. Throughout this book, such reality is mixed with fantasy, requiring the reader to continually distinguish between the two. This process will hopefully promote thought and learning, for as a wise old cat once said: "To learn, it helps to think." With this in mind . . .

1. CATS AND THE SUN

Cats love the sun. Some think they invented the sun. While modern science has been unable to prove this theory, modern science has been just as unable to explain gravity. Nevertheless, no one can dispute that cats have enjoyed a long and colorful association with the sun, stretching back at least to the dawn of recorded human history.

Figure 1.1 Cat scientist inventing the sun.

1.1 HISTORY

It is well known that cats were worshiped in ancient Egypt. The Egyptians recognized how perfectly cats interact with their environment and glorified them appropriately. In ancient Egyptian art the goddess Bast represented the life-giving heat of the sun and was portrayed with the head of a cat. Quite obviously those crafty Egyptians recognized the close connection between cats and the sun. This "Egyptian Period" witnessed tremendous growth of solar knowledge by cats. During this epoch cats discovered thermal mass.[1]

Figure 1.2 Ancient Egyptian cat discovering how the tremendous thermal mass (or heat storage capacity) of the pyramids provides unchanging cool relief from the scorching desert environment.

Even after the decline of Egypt, cats continued to advance in their use of the sun, though at times their progress was slow. They traveled to Europe where their solar research activities were severely hampered with the onset of cat-hating religious cults of the Middle Ages. Cruel and arrogant humans forced cats to stay outside where they became experts at chasing mice and rats for survival and entertainment.

People have since turned from coal to oil and natural gas as their primary energy sources. This change from messiness to sliminess and smelliness has long been viewed as part of a logical progression first to nuclear fission,[2] and then to nuclear fusion.[3]

---

1. Thermal mass is anything that holds heat and especially things which hold a lot of heat.

2. Nuclear fission is lots of atoms splitting apart making heat energy and radioactive waste material.

3. Nuclear fusion is where even more lots of energy is released when lots of atoms are fused together. Humans have not found a way to use nuclear fusion energy on earth in any practical sense, except in hydrogen bombs which make practically no sense at all.

Figure 1.3 Although medieval European winters were very cold, cats learned to use the sun to stay warm, and thus survived this era of persecution. By the early eighteenth century cats had regained human favor, having eliminated the rats who carried the Black Plague throughout Europe.

Figure 1.4 At the onset of the industrial revolution cats were much more highly regarded, but with cheap and plentiful coal it seemed hardly worth the effort to study such a seemingly weak energy source as cats using the sun. Being basically messy, people preferred filthy coal to self-cleaning solar cats.

Unfortunately,[4] nuclear fission cannot be developed as quickly as once thought. Dogged and contentious debate over whether the dangerous radioactive materials produced by this technology can be kept from the living environment may well continue for as long as the potency of the materials in question.

And as for fusion energy, cats already solved that problem. By inventing the sun and locating it a safe, yet convenient distance from earth, cats have arranged for large amounts[5] of solar energy to be delivered to earth each day at no cost whatsoever! So the great, inevitable shift to solar energy has begun. Starting with President Carter's remarkably cat-like response to the first modern "energy crisis" of the 1970s, peoples' acceptance and support of solar energy has grown steadily.

Starting with President Carter's strong commitment to fund alternative energy research and development in the late 1970s, an unbroken string of U.S. presidents have conducted a relentless campaign in support of solar energy development.[6]

Figure 1.5 Humans engaged in the Great Nuclear Debate.

4. Or fortunately.

5. The sun is just a really big fusion reactor, which in the year 2000, delivered over 13,000 times more solar energy to the earth than was used by all of human civilization for all purposes that year.

6. Since all "fossil" fuels including coal, oil, and natural gas are nothing but solar energy stored in chemical form, it is possible to think that the U.S. has led the world in using solar energy at ever increasing rates. It is not clear however, what good such thinking actually does.

Figure 1.6 President Jimmy Carter proclaiming that the Energy Crisis of the late 1970s was M.E.O.W. (the Moral Equivalent Of War).

Figure 1.7 President Ronald Reagan cleverly redefined the energy crisis as nothing but A.R.F. (Another Regulatory Fiasco), while appearing to do nothing to support solar energy.

Figure 1.8 President George H. Bush appeared to subvert Carter's famous phrase by insisting that the 1991 Gulf War was about freedom and was not the first M.E.O.W. (Middle East Oil War) while he actually raised renewable energy research to national laboratory status.

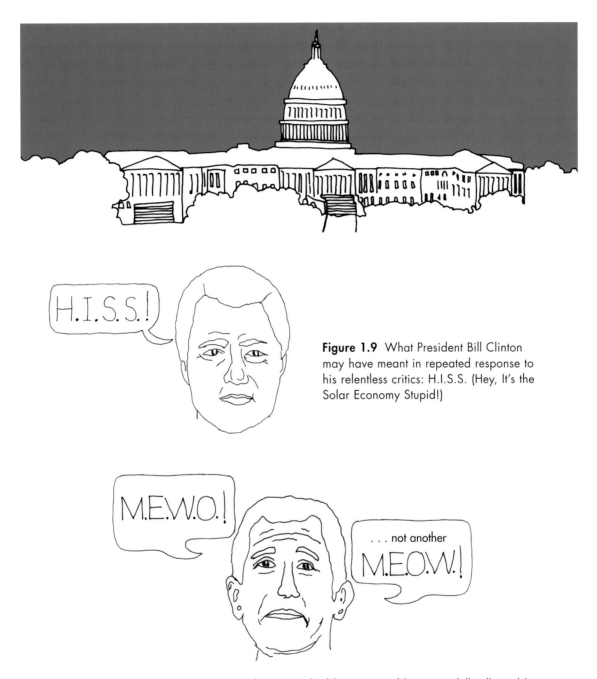

Figure 1.9 What President Bill Clinton may have meant in repeated response to his relentless critics: H.I.S.S. (Hey, It's the Solar Economy Stupid!)

Figure 1.10 President George W. Bush's unmatched business and linguistic skills allowed him to define his energy policy to be M.E.W.O. (More Energy With Oil), while compassionately following in his father's missteps.

Cats know a lot more about solar energy than people do. This sizeable solar knowledge gap will narrow as more and more people come to realize that the sun is their best and is really their ultimate energy source, and begin to devise ever more ingenious ways of using it to their own advantage. Encouraging evidence of just such a trend can be seen by carefully measuring and comparing solar knowledge over recent decades. In 1980, as the shock of the first "energy crisis" took hold, people began to try very hard to learn about solar energy, though they were very far behind cats. But by the turn of the twentieth century a detectable increase in human solar knowledge took place, although humans still have a long way to go to catch up to cats.

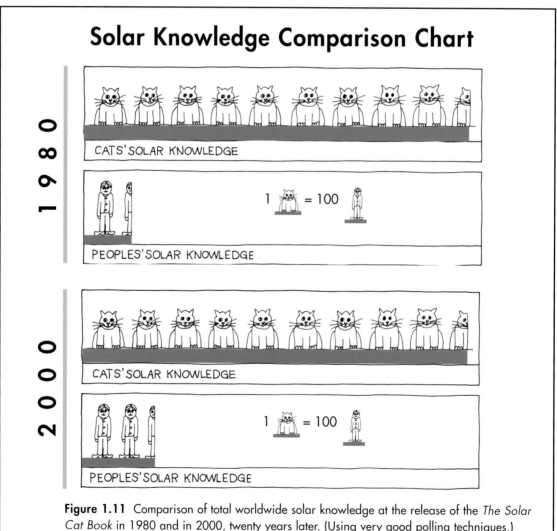

Figure 1.11 Comparison of total worldwide solar knowledge at the release of the *The Solar Cat Book* in 1980 and in 2000, twenty years later. (Using very good polling techniques.)

Today, cats can be seen routinely using the sun to meet their energy needs. How common is the sight of a cat curled up in a patch of sunlight? A good guess would be 140 million times a day in the United States alone! This is a good guess because there are about 70 million cats in the U.S.,[7] and if on average, each cat is seen twice per day, 140 million daily sightings result.

Figure 1.12 Two sightings of a cat curled up in a patch of sunlight.

---

7. *U.S. Pet Ownership and Demographic Sourcebook,* published in 2002 by the American Veterinary Medical Association's (AVMA) Center for Information Management.

1.2 THE STATE OF THE ART

In spite of their recent progress, many people are still reluctant to admit that cats know more about solar energy than people do. At times it seems that people refuse to see even the most obvious. A common example of this involves cats' affinity for sunlight on chilly afternoons. Generations of study have led cats to the astute conclusion that when cold, it makes perfect sense to expose one's fur to the sun. The human approach has not always been so direct.

Figure 1.13 Cat solving a problem in 2 steps.

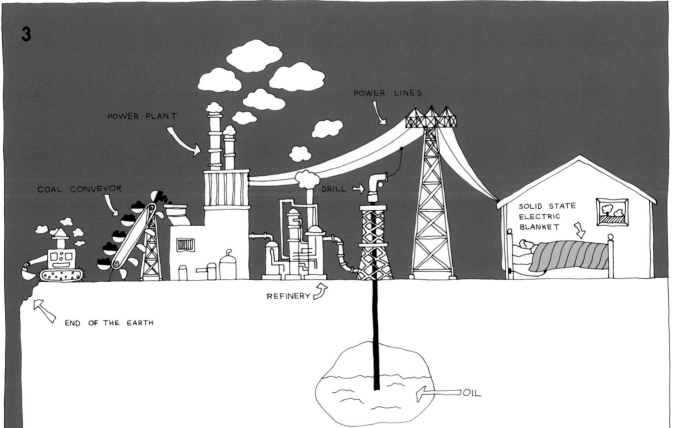

Figure 1.14 Person solving the same problem in more than two steps.

Cats not only use the sun in their everyday lives, they also work hard to improve and per-fect the ways they use solar energy. At times, however, it is not so obvious that cats are involved in solar research and development. Their well-known fondness for going outside in the dark is a prime example of this. Common folklore has it that cats can see in the dark and can therefore hunt well at night.[8] Naturally, this has never been proven since people can't see in the dark to tell what cats are doing out there.

Figure 1.15 Imagined view of a cat in the dark looking for the sun.

8. People have made great progress in recent years learning how valuable sunlight can be in providing high quality light to the inside of buildings. They call this "daylighting." People have also made even more astounding gains in producing efficient and aesthetic artificial lighting products. However, they have yet to even consider the possibility of adapting their eyes (as cats have) to a wider range of lighting conditions.

Bird chasing is another well-known facet of cat behavior that has been mistakenly attributed to simple "animal instinct." Their real motive in such pursuit is to learn how birds fly through capture and thorough inspection. Once cats learn how birds fly, they can then get closer to the sun on very cold days.[9]

Figure 1.16 Flying cat getting closer to the sun.

9. The first release of this book brought cries of dismay from many people at how such seemingly ridiculous ideas as this could possibly be true. But that was before the introduction of the U.S. Government's "Clean Coal" Initiative.

It is simple to see how cats get warm by exposure to sunlight: when it hits them, they heat up! Of equal importance are ways they avoid the sun and the reasons they do this. Obviously, the primary reason cats avoid the sun is to stay cool.

Figure 1.17 Cat displaying the ability to recognize when it is too hot and to do something about it.

Figure 1.18 Another way that cats avoid the sun on hot days is by crawling into paper bags. This displays their use of hide technology.

Sometimes cats are involved in solar research and development activities when they appear to be just standing around. Many of their research projects deal with such subtle forces that people can barely conceive the true purpose of such apparently ordinary behavior.

Figure 1.19 Cat absorbing squiggly, straight, and dashed heat.

Cats know a good thing when they see one, be it a fresh, juicy fish or an efficient energy source. People are equally perceptive regarding fish, but have some trouble recognizing good energy options. The remainder of this book is more about energy than fish to help people strike a better balance.

2. SOLAR THERMAL CATS

Solar thermal cats convert sunlight into heat, and are similar in many ways to peoples' solar thermal systems. This chapter describes their basic types and component parts offering frequent comparison to peoples' systems.

2.1 PASSIVE CATS

There are two kinds of solar thermal cats: passive cats and active cats. Passive cats sit around all day, often on window sills, absorbing sun and later letting it go. They hardly do anything except to occasionally go to their food or water dish, or perhaps to their litter box.

Figure 2.1 Passive cat.

2.2 ACTIVE CATS

Active cats, on the other hand, get up and go after the sun, be it on a garage roof, or on the back fence. They then bring it back inside where it will do some good. All this running around makes active cats much bigger eaters than passive cats, so active cats generally cost more to own and operate. Solar thermal systems designed by people are much the same but are unable to change from active to passive (or vice versa) as cats are known to do.[1]

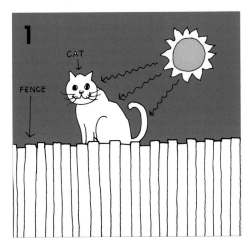

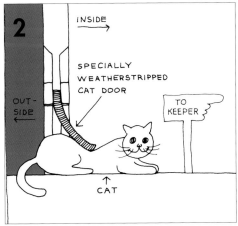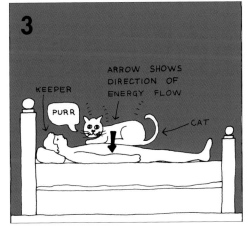

Figure 2.2 Active cat, Modes 1–3.

1. The most notable distinction between peoples' passive and active systems is that their active systems need electricity to work while their passive systems do not.

To fully appreciate the elegant simplicity of solar thermal cats it is useful to first study the different parts of peoples' systems. Whether passive or active, peoples' systems have four basic parts: collector, storage, heat transfer mechanism, and control.

Collector   All solar thermal systems collect energy during the day, heating the home or office as necessary. In passive systems, collectors take the form of windows, skylights, and greenhouses. In active systems common flat plate collectors are used.

Storage   Once daytime heating needs are met, additional energy is stored for use at night. Such storage can be in tanks of water, piles of rock or dirt, or almost anything else that can hold heat. In passive systems, heat storage often occurs in the building structure itself, as in concrete or tile floors, thick plaster or wallboard walls, or even stone or cast metal furniture.[2]

Heat Transfer Mechanism   In addition to collectors and storage, solar thermal systems must have some way to transfer heat from the collector to the point of use. In active systems, pumps, fans, and a "heat transfer fluid" are used. In passive designs, the heat flows all by itself, such as when warm air rises or heat seeps through a wall.

Control   In active systems, electronic controls tell pumps and fans when to run in order to collect, store and distribute energy. In passive systems this control function occurs automatically, though it is sometimes necessary for a person to open and close windows, curtains or other such controls.[3]

---

2. Even though such furniture may be rare, it holds heat pretty well.

3. While the complete elimination of the need for human response to or even awareness of the natural environment has been a guiding principle of modern design, it has been shown that such activity may in fact foster health as well as happiness, while providing a degree of beneficial cardiovascular and psychological exercise.

With cats, all system functions are combined into a single, nearly flawless component: THE CAT. In a solar thermal cat, the cat serves as collector, storage, heat transfer mechanism, as well as system controller. In addition, cats make good use of catabolic heat generation,[4] an aspect entirely missing from peoples' systems.

The following sections describe each of the basic solar thermal system components in greater detail.

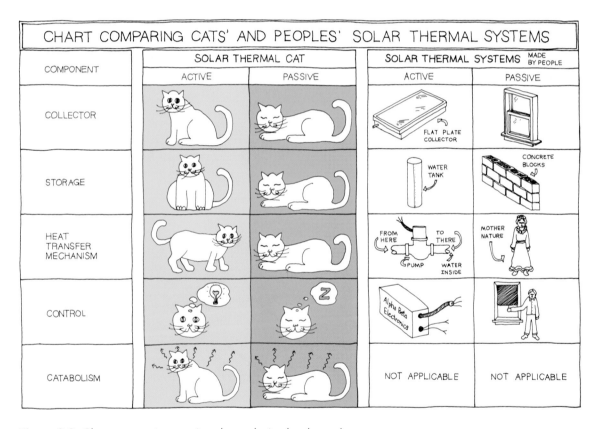

Figure 2.3 Chart comparing cats' and peoples' solar thermal systems.

4. Catabolism is the part of animal body metabolism in which heat is released. Anabolism is the other part.

2.3 CATS AS COLLECTORS

Cats have several inherent features which allow them to function as highly efficient, aesthetic solar collectors. Among these are the "heat trap" characteristic of cat fur, cats' ability to be concentrating, tracking, or fixed, and their unique self-cleaning characteristic. Peoples' collectors have many of the same features but are not usually self-cleaning.

Heat Trap Fur Cats' high solar collection efficiency is primarily due to the "heat trap" characteristic of their fur. When cat fur is properly oriented, squiggly heat rays from the sun neatly slip between the fur to strike the cat skin underneath. This occurs regardless of fur color, but is affected by fur length, thickness and straightness.

Since cat fur prevents wind from getting near the heated cat skin, solar energy striking the cat body is trapped, hence the name "heat trap fur."

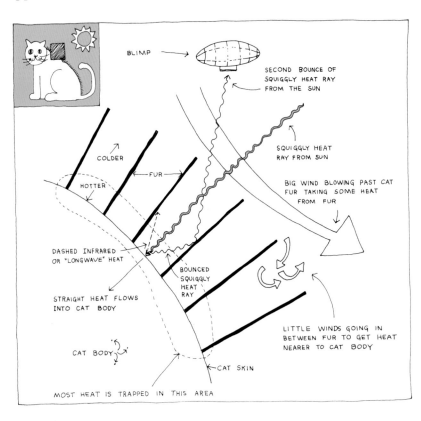

Figure 2.4 Energy flow diagram showing the "heat trap" characteristic of cat fur in a properly oriented cat.

People's Solar Thermal Collectors While some people have tried to imitate cat fur's "heat trap" characteristic, most still revert to using glass or plastic "glazings"[5] and special coatings to reduce heat loss to the air from the collector's absorbing surface. In most locations a single glazing works fine for collectors in solar water heating systems. Some collectors use two glazings with an air space between them to reduce heat loss to the surrounding air. The most common way to reduce collector heat loss however, is to use a single glazing and a special coating[6] on the absorber surface that minimizes heat loss by radiation.

Even unglazed collectors (similar to clean-shaven cats) can be effective. They can be used for swimming pool heating in cases where the collector absorber is very nearly the same temperature as the surrounding air, so little heat loss results.

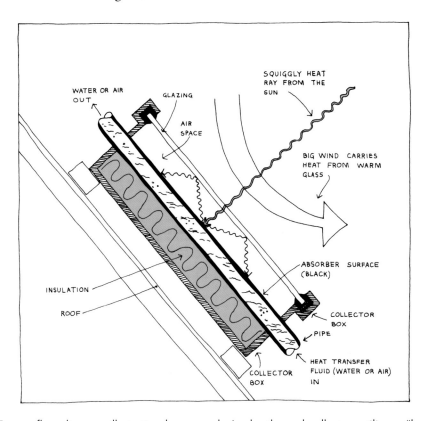

Figure 2.5 Energy flow diagram illustrating how peoples' solar thermal collectors utilize a "heat trap" principle of their own, but without fur.

5. Glazing is an architectural term for transparent materials which can be used for windows or flat plate collector covers.

6. Called "selective surface coatings" the most common is known as "black chrome."

Catacombs Since fur orientation is critical to solar cat collection efficiency, methods to ensure proper orientation are extremely important. One promising way of doing this involves combining a cat door with a fur comb so that whenever a cat goes through the door (or catacomb) its fur is properly arranged. No good method has yet been devised to insure that the cat positions itself properly once it passes through the door but science and technology will surely solve this problem in the end.

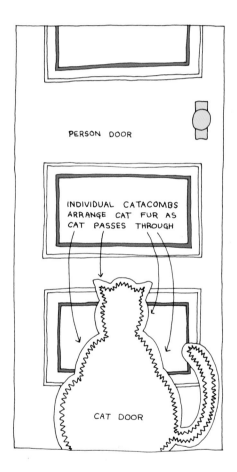

Figure 2.6 Special catacomb cat door designed to arrange fur in a desired manner as cats pass though it.

Concentrating Cats   Concentrating cats are quite common. They have been known to stare intently for hours at what appears to be nothing. No plausible explanation for this has yet been advanced, though many people suspect some sort of intense energy collection or channeling technique is in play. Peoples' concentrating collectors focus or concentrate the solar energy from a large area onto a much smaller area, just like a magnifying glass in the sun. They usually need to stay facing the sun as it moves across the sky.

Figure 2.7 Cat concentrating on matters far beyond human comprehension.

Fixed Cats and Tracking Cats   It is important that fixed cats are not confused with peoples' "fixed" solar collectors. The same is true for tracking cats and peoples' tracking solar collectors. There are two kinds of fixed cats. The first kind are fixed to limit the cat population and in the case of boy cats, to suppress aggressive behavior. The second kind are fixed because broken cats just don't work right.

Figure 2.8 Broken Cat.

Figure 2.9 Fixed Cat.

Peoples' fixed collectors are those which stay in one place and do not change their tilt or orientation to follow the sun across the sky. Tracking cats are those which sniff out and locate patches of sunlight, birds or mice. Peoples' tracking collectors change their tilt and orientation to remain faced directly at the sun all day long, and have very little to do with birds or mice.

2.4 CATS AS THERMAL STORAGE

There are two kinds of thermal storage:[7] "sensible" and "latent." Cats can store heat in both ways, whereas peoples' solar thermal systems usually use only one or the other.

Sensible Heat Storage Sensible heat storage is "sensible" because it is easily sensed. When something stores sensible heat its temperature increases. Heavy things store more of it than lightweight things, with the same temperature rise and size of thing. Also, a lightweight object must get much hotter (in temperature) to absorb the same amount of sensible heat as something very heavy. Cats are so good at storing sensible heat that a basic measure of heat energy (the "CTU" or Cat Thermal Unit) has been named after them.[8]

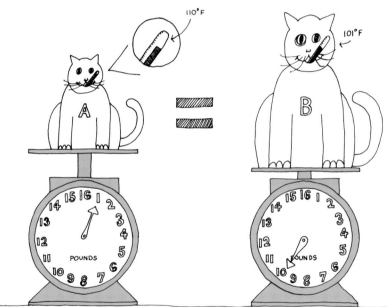

Figure 2.10 Both cats A and B started out having temperatures of exactly 100°F. and then proceeded to absorb the same amount of heat. Cat A became much hotter than cat B because cat B weighs so much more.

---

7. "Thermal storage" is just a fancy name for heat storage.
8. CTUs and heat are explained more fully in the Appendix.

Latent Heat Storage   Latent heat storage is not nearly as obvious as the sensible kind. This is because the temperature of the object (or cat) does not change as latent heat is stored or removed. People have experimented with latent heat storage in solar thermal systems and in building designs by using special "phase change" materials. which store or release heat by melting or freezing at convenient temperatures like room temperature.

With cats, latent heat storage is not so tricky. It takes place in the formation of purrs. When a cat has been warmed to its normal limit of 102°F excess energy can be stored in what are called "latent purrs" The exact amount of energy that can be stored in a latent purr is unknown. However, some of the softest purrs have warmed the coldest hearts, and it would be folly to ignore such startling effects.

Figure 2.11 The effect of a latent purr on a cold heart.

The Insulating Value of Fur A common fault of many of peoples' solar thermal systems is that their storage tanks are not well insulated. This is usually not the case with cats. Cat fur is widely recognized as perhaps the ultimate insulating material.[9] It adjusts its insulating value to follow changing seasonal conditions (as in shedding), and grows back to repair shaved or damaged areas.

People may be able to imitate the appearance of cat fur in their own insulating materials. However, it is doubtful that they will ever duplicate the self-adjusting and repairing aspects.

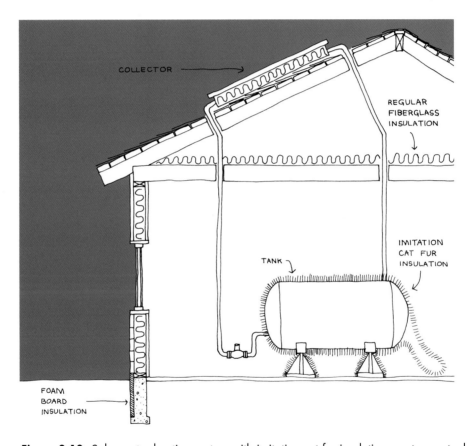

Figure 2.12 Solar water heating system with imitation cat fur insulation on storage tank.

9. While some other materials, like polar bear fur, may actually exhibit higher insulating value than cat fur, it is just not practical to let such animals have the run of your house.

2.5 CATS AS HEAT TRANSFER MECHANISM

It is easy to see how cats transport heat. They simply get up and take it wherever they go. People have not yet developed as good a way to move heat around, except perhaps in passive systems where the heat moves around more or less by itself. In active systems, people use liquids or air to transfer heat. Such fluids can leak, turn corrosive, or freeze if care is not exercised.

Figure 2.13 Cat collecting heat just prior to transporting heat.

Figure 2.14 Cat having transported heat showing the distance transported.

Freeze Protection Cats avoid freezing by wearing fur coats and by staying in warm places. Peoples' solar thermal collectors have a much harder time since they spend their entire lives outside without hope of comfort on cold nights. This doesn't pose a great problem except for those collectors which contain liquids which can freeze, possibly damaging the fluid passages.

Fleas Protection While freeze protection is a critical issue in peoples' solar thermal systems, fleas protection is of equal concern to solar thermal cats and their owners. Though there is much interest in finding better ways to discourage fleas from bumming around in cat fur, no outstanding method of fleas protection yet exists. Many flea sprays, collars, tags, and pills containing chemicals or anti-flea bio-agents are available; however, a weekly bath remains the safest, most natural, and effective method since fleas enjoy swimming even less than cats.

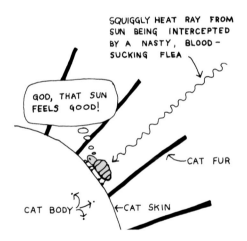

Figure 2.15 Diagram of a flea in a working position.

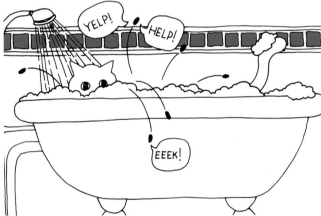

Figure 2.16 Fleas jumping from bathing cat in "fleas protection" mode.

Corrosion Cats do not corrode. However, the metallic parts of flat plate and concentrating solar thermal collectors can, and things must be arranged to keep this from happening. This is usually accomplished by avoiding the use of corrosive materials or by using chemical additives to protect the metal surfaces from corrosion.

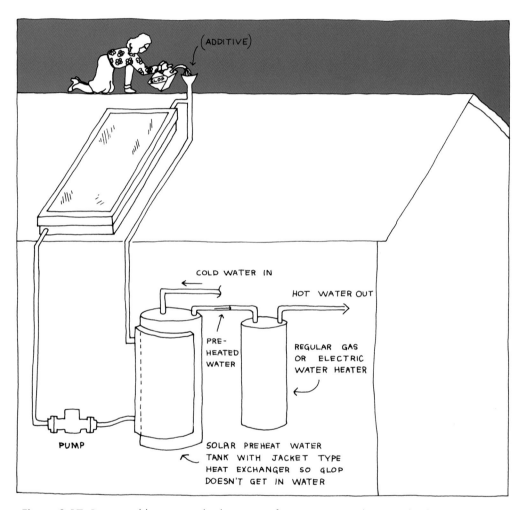

Figure 2.17 Person adding prescribed amount of anti-corrosive glop to solar heating system collector "loop" at prescribed interval.

With solar cats, the cat itself is the corrosive agent. Some cats take great pleasure in sharpening their claws on common household objects. Prevention of such accelerated wear of internal home surfaces can also be accomplished with the use of additives. Adding a piece of furniture which begs to be scratched is a good additive.

Of course, another method to prevent such damage is to avoid the use of cats, but this is clearly out of the question.

Figure 2.18 Person adding funny looking scratching post to inside of cat's house to prevent corrosion of other structures.

Overheating A common problem with flat plate or concentrating solar collectors is overheating ("stagnating" in solar jargon). On sunny, hot days when the last thing anyone wants is solar heat, collectors can get extremely hot, sometimes causing heat transfer fluids to turn corrosive or causing other kinds of material failures.

Overheating is never a problem with cats since they move to shady places when the sun gets too hot for them. Overeating is a problem for them though, similar in some ways to alcoholism in people.

Figure 2.19 Cat overeating.

2.6 CATS AS SYSTEM CONTROLLERS

Cats are always in control. They know exactly what to do and when to do it. The only possible explanation for this is that they each possess an incredibly sophisticated and complex (although very tiny) control network composed of a constellation of very powerful computers much like that of a large power plant.

Peoples' solar system controls are not nearly as complicated or socially significant. Active systems use relatively simple electronic control units. They sense temperatures in various places (principally at the collector and storage tank) and tell other electric components what to do at the proper time. Passive systems are either self-regulating by their very design, or require people to do things like adjust shades, or open or close windows.

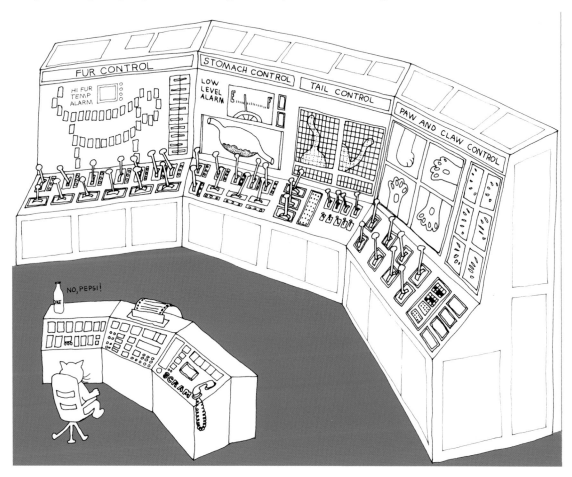

Figure 2.20 Very enlarged view of a cat brain central control room.

2.7 CATABOLISM

Catabolism is the part of body metabolism in which heat energy is released. The catabolic heat generation rate of a cat depends on its state of activity. For example, a cat at rest discharges catabolic heat at a rate proportional to its weight, approximately according to the following equation:[10]

$$Q = 16.5 \ W^{0.75}$$

Where: Q =   the catabolic heat generation rate in BTUs per hour per cat

W =   weight of the cat in pounds

From this equation, the standard 10 pound cat[11] would produce 92.8 BTUs per hour, enough to raise a quart of water from 55° to 100°F.

This is a rather small rate of release, especially when compared to the current total U.S. energy use rate of 2.94 billion BTUs per second. However, even this gargantuan energy demand could be met with the catabolic heat generated by a single, 5.39 million billion pound cat, or only 70,449 100–megaton cats as described in Chapter Three. However, there is some question as to whether or not such large cats will ever be developed.

It is certainly possible to breed enough cats of normal weight to satisfy U.S. energy consumption with catabolic heat alone. About 117.1 billion "standard" cats would be adequate for this task. The current U.S, cat population is about 70 million. Luckily, cats are very fast breeders, and like their nuclear counterpart, can theoretically multiply fast enough to meet the challenge.

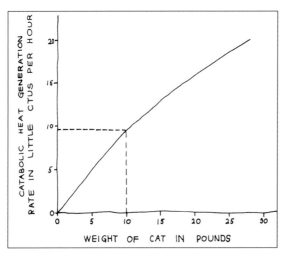

Figure 2.21 Graph of catabolic heat generation rate versus weight for normal weight cats. Both Big and Little CTUs are defined in the Appendix.

---

10. Reprinted with permission from the *1977 Fundamentals Volume ASHRAE Handbook & Product Directory.*

11. The "standard cat" is defined in the Appendix.

Exactly how quickly cats could provide the world with catabolic energy self-sufficiency is easily determined with some simple arithmetic. If the following assumptions are made, approximately 6.1 years[12] would be needed to "fast breed" the necessary number of cats:

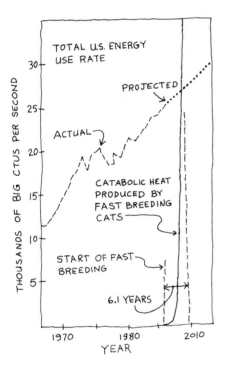

1. Everyone tries very hard to help cats breed.

2. The starting cat population is 70 million.

3. Half of all cats are female.

4. Half of all female cats at the commencement of fast breeding are fertile.

5. Each fertile female cat can have two litters of four kittens each year.

6. Female cats can litter only between the ages of one and ten.

7. The average life span of all cats is 15 years.

8. All cats are "standard" cats.

Figure 2.22 Chart showing how quickly fast breeding cats could solve the energy crisis with catabolic heat generation alone.

12. In the 1979 edition of The Solar Cat Book, this time was 9.3 years due in large part to the fact that at that time, the starting cat population was only 59 million cats. This leads to the interesting possibility that if people wait long enough, the solution to their energy problems may become easier without their doing anything at all!

2.8 SOLAR THERMAL APPLICATIONS

The normal function of solar thermal cats is to keep cats warm. However, they can be used by people in a variety of ways. The following examples illustrate a few of the ways solar thermal cats can be applied to human needs.

Residential Residential applications of solar thermal cats are most practical since cats like to stay close to home. Solar thermal cats are best at space heating but can heat domestic water, swimming pools, and even hot tubs, under the right circumstances.

Peoples' active solar thermal systems are most practical when attempting to supply low temperature heat. Swimming pool heating is therefore most effective, followed by hot tub, domestic water, and then building space heating. Passive space heating systems are much more practical than active ones, since they cost less and work much better.

Figure 2.23 Solar thermal cat pre-heating a hot tub in a residential setting.

Commercial Solar thermal cats can also be used in commercial settings, especially in restaurants with tasty table scraps. Peoples' commercial systems have principally been used for water heating in restaurants and hotels, and for swimming pool heating.

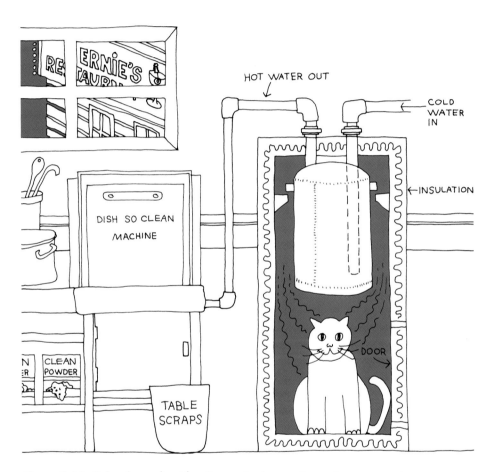

Figure 2.24 Solar thermal cat heating water in a restaurant.

Industrial People have been slow to adopt solar thermal systems in industry. This is partly because the relatively low cost of industrial energy fostered great waste. In recent years, industry has found investment in energy conservation much more profitable than in solar thermal energy production, but is expected to slowly embrace solar technology.

Cats understandably avoid smelly, noisy factories. They can, however, be persuaded to provide people with heat for industrial processes if the rewards are equal to the task.

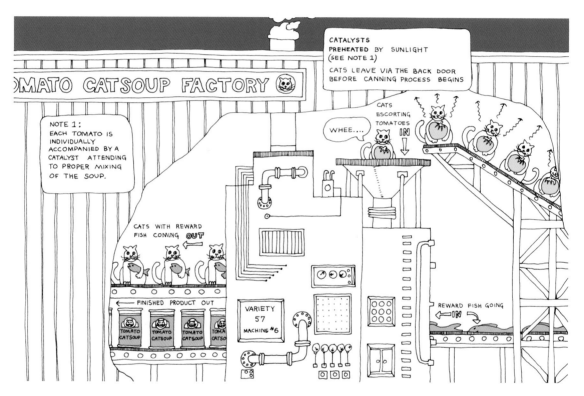

Figure 2.25 Warm solar thermal cats reduce process heating requirements in tomato cat soup factory.

There are few solar electric cats in the world today. This is because cats have little use for electricity. While they do like to sit atop warm television sets and electric blankets, they see little sense in expending the effort required to produce electricity for heating when they can much more easily get warm by just sitting in the sun. This is not to say that cats see no value in electricity. They are strongly in favor of maintaining sufficient electric power generating capacity for certain of their needs.[1] However, as long as people kindly supply them with free food and electricity, cats will keep their own power generating capacity in reserve—for reasons that will become apparent later in this chapter.

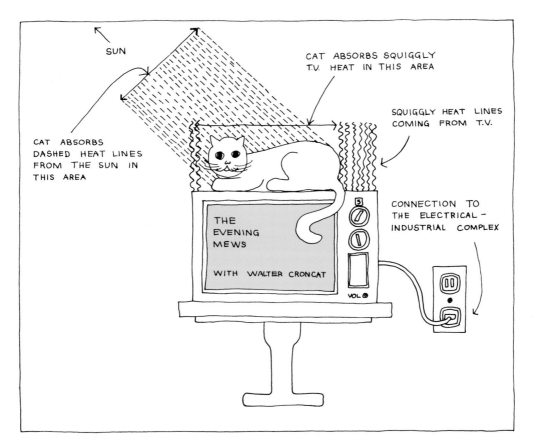

Figure 3.1 Cat appreciating the free electricity that humans so graciously provide, while being mindful that it's a whole lot easier to get warm using direct solar energy.

---

1. These include cat food and cat toy production, cat health care services, and catland security.

3.1 THE IMPORTANCE OF ELECTRICITY

People, on the other hand, can barely conceive of life without electricity. When it was first introduced, they were astonished at its seemingly unlimited potential. Today electricity is as commonplace as background radiation, and is used in virtually every aspect of modern life.

Peoples' great demand for electricity has caused them to study and develop many methods of power generation. Among these are several promising and a few not so promising ways of converting sunlight into electricity. The most important of these methods are described in the remainder of this chapter, as well as the method cats discovered as the best way to convert sunlight to electricity. Keep in mind when reading this chapter, that not everything people do necessarily makes sense.

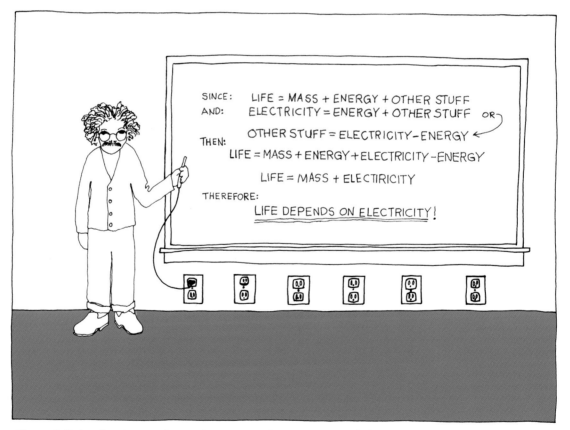

Figure 3.2 Leading human theorist lecturing on the importance of electricity to life.

Electrical Energy Conservation People usually engage in energy conversation when they would be much better off engaging in energy conservation. This most often concerns gasoline but it also applies to electricity. It may be a simple matter of the two words being so similar that people just can't tell the difference.[2] Cats were once confronted with such a problem. When they first learned to speak nearly everything they said sounded like "meow."[3] They overcame this problem by learning how to listen carefully. It may not be feasible to expect people to listen carefully, but other ways may be found to help them.

Figure 3.3 Sophisticated psychological experiments now being conducted to find the best way to teach people how to conserve energy

2. It may also be due to a basic confusion between the meanings of virtue and necessity.

3. Nearly everything they say still sounds like "meow" to people at least, but cats know the difference.

3.2 WIND ELECTRIC GENERATORS

Cats usually do not like wind[4] because it messes up their fur and makes it difficult for them to hear birds and approaching dogs. However, they do agree that the wind carries a great deal of energy and would surely harness it if their electrical needs were greater.

People have long applied wind energy to their own needs. Windmills have been used throughout the world for centuries, for milling food and pumping water.

More recently, people have found wind electric generating machines to be the easiest way to convert solar energy into electricity.[5] People have become so good at this, that giant wind "farms" have been built all over the world, and wind power has become as inexpensive as conventional fossil fueled power plants in many places.

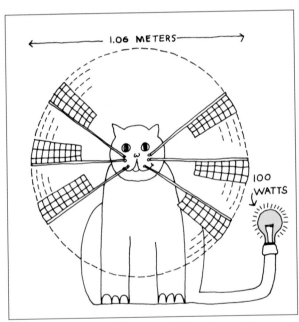

Figure 3.4 A very efficient wind electric catmill as seen by a 15 mph wind.

But in their insatiable drive to grow at ever increasing rates, humans in the wind electric industry have begun to incorporate cat-like features in new and exotic wind machines. It is hoped that by doing so, efficiencies can be increased, and other impediments to wind energy development can be overcome.

Of course higher efficiencies will mean fewer machines need be installed to produce the same amount of energy. This would reduce costs, thereby further increasing its competitiveness. In addition, the value of the federal purrduction cats credit (known as the "PCC") would increase since catmills would keep purring even if the wind stopped!

---

4. Many cats actually love the wind, especially when it creates opportunities to suddenly chase things that don't seem to be there!

5. Wind is caused by uneven heating of the earth's surface and atmosphere by the sun, so wind energy is really an indirect form of solar energy.

Fewer, more efficient machines will also help dispel what is perhaps the most problematic objection to wind energy development: aesthetics. Some people object to the very idea of even nice looking wind turbines dotting beautiful landscapes.[6] Not only would the sight of large, graceful cats dramatically improve nearly any vista, use of familiar markings[7] could make them nearly invisible! Cat-like features could also eliminate objections to wind-turbine noise![8] Imagine the pleasure of hearing a sweet chorus of whispurring cat-mills!

Harm to bird life has also been cited as a serious consequence of wind turbine operation. People have effectively eliminated this problem by avoiding locations that are in migratory flight corridors for birds.[9] Making wind turbines look like cats would be an excellent solution to this problem as well! Not even the most powerful bird would dare fly anywhere near a bunch of 70-meter tall cats!

Figure 3.5 Imagined view of a large scale Wind Electric Catmill Farm.

6. This is an unavoidable problem in the same way it is for coal, gas and nuclear power plants—they are all big enough to be seen, and there will always be people who just don't like the look of them!

7. Tiger stripes, jaguar or leopard spots, for example.

8. People have reduced the sound of these machines so much that they often cannot be heard above the sound of the wind in the grass.

9. The birds were there first, after all.

The Power in the Wind One of the more interesting things about wind is that the power in it increases very rapidly[10] with its speed. For example, a 20 mph wind has eight times as much power as a 10 mph wind. However, no matter how fast the wind blows, people have not found a way to extract all its energy.[11] This seems to make sense since the energy in wind comes from its speed, so to extract all its energy you would need to remove all its speed! The air would just pile up at the point of extraction since it would have no speed to go anywhere else!

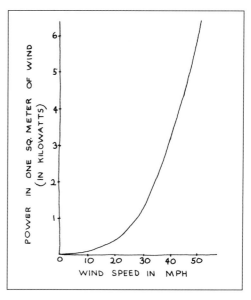

Figure 3.6 How the power in wind increases with its speed

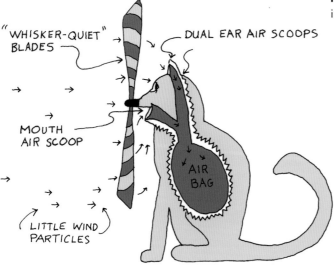

Figure 3.7 Cutaway view of an experimental, 100% efficient wind electric catmill.

10. Scientists and even wind energy enthusiasts often use the word "geometric" or sometimes "exponential" to emphasize this fact and to sound really smart.

11. People are largely resigned to the notion of a maximum theoretical limit of about 59% of the energy in wind can be extracted for any use.

But even as their modern wind turbines have nearly reached this limit, the most aggressive turbine makers insist on attempting to break this barrier! Their most plausible idea so far has been to incorporate giant internal wind bags into which all that piled up air could be kept until the real wind died down at which time even more energy could be extracted as the catmill exhaled!

Figure 3.8 This somewhat large, 57% efficient wind electric catmill shown against the skyline of Chicago, would produce 30 megawatts in the Windy City's 10.2 average wind speed.

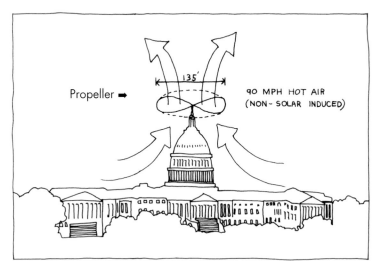

Figure 3.9 This markedly smaller wind generator mounted atop the U.S. Capitol Building would also produce 30 megawatts because the hot air velocity near this landmark is so high.

3.3 PARABOLIC TROUGH SOLAR THERMAL POWER PLANTS

The largest scale and most efficient solar electric technology people have so far managed to develop is called the "parabolic trough solar power plant." Such plants consist of long rows of mirrors curved in the shape of parabolas, which bounce and focus sunlight onto insulated pipes, heating a fluid to make steam and drive a turbine just like in a good old fossil fuel burning power plant.[12]

This technology works best in nice hot places with lots of sunlight, like deserts, and at very large scales, just like with good old fossil fuel burning power plants. The only thing they can't do nearly as well as good old fossil fuel burning power plants is to burn fossil fuels.

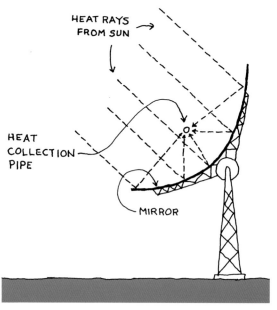

Figure 3.10 Parabola-shaped mirrors reflect sunlight onto horizontal pipes.

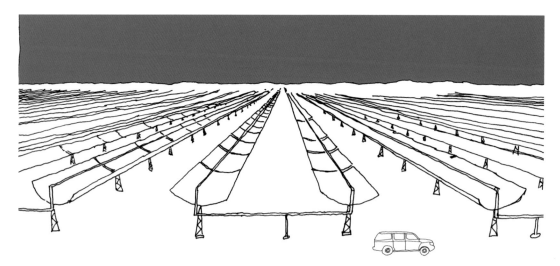

Figure 3.11 Portion of a large "parabolic trough" solar power plant in the desert with an object of familiar size.

12. Sometimes engineers and scientists refer to this as a "Rankine-cycle" power plant.

Nine plants were built in California's Mojave desert in the late 1980s and early 1990s. Those plants produced more electricity through the end of the last century than all other solar electric technologies[13] in the entire world combined! In spite of this good record, and real potential[14] for this technology, people have found it nearly impossible to set consistent laws and policies to develop this proven technology. This may be due to an inability of their policy makers to clear up their confusion with a more familiar kind of trough.[15]

Figure 3.12 Policy makers of differing breeds agreeing on one point while missing another.

13. This includes all direct solar to electricity conversion technologies including power towers, dish sterling systems, and photovoltaics, but does not include wind or hydroelectric energy.

14. Filling one fourth of the Mojave with parabolic trough solar power plants could supply the entire U.S. annual electricity demand of about 3,700,000,000,000 kilowatt hours. There are many other bigger deserts in the U.S. and elsewhere, so land area is not a problem.

15. Another explanation is that people are distracted by energy buzz-words like "distributed generation" and "hydrogen economy," forgetting that as long as there are cities, centralized power generation and distribution won't go away, and the need for reliable, clean electric power will remain essential.

3.4 POWER TOWER SOLAR ELECTRIC PLANTS

Another large-scale solar electric technology called "Power Tower" plants have been the focus of much government attention. Some have even been built as U.S. government research projects. These plants have acres of computer controlled mirrors, each of which bounce sunlight onto a small receiver atop a tall central tower. A great amount of sunlight can then be concentrated, producing very high temperatures to make steam and electricity with conventional turbine generators.

These research projects have been very costly, never approaching the simplicity and effectiveness of parabolic trough plants. Many in and out of government have come to see these plants as resembling giant turkeys.

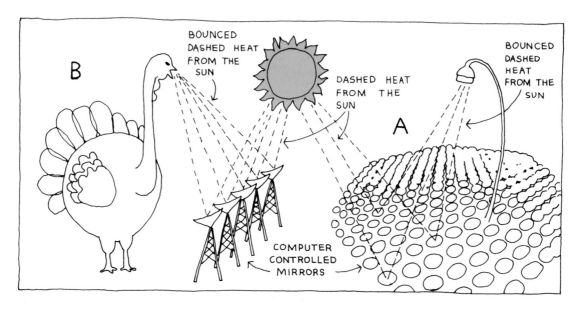

Figure 3.13 Two views of Power Tower solar electric technology: A. how it might actually work, and B. What government contractors have so far managed to achieve.

To avoid problems with turkeys, the U.S. Department of Energy, in cooperation with the Department of Defense, started construction of a giant solar cat in the desert near Barstow, California. The secret, $1.6 trillion project would have produced a 100 million ton cat which mimicked a solar cat's basic method for capturing energy on a grand scale.

In addition to simple power generation, government planners had hoped to teach this cat to catch Soviet ICBMs if they were birds, and scoop enemy submarines from the oceans as if they were fish. But as the Soviet threat receded, and cheap oil again flooded world markets, support for this exciting project vanished.

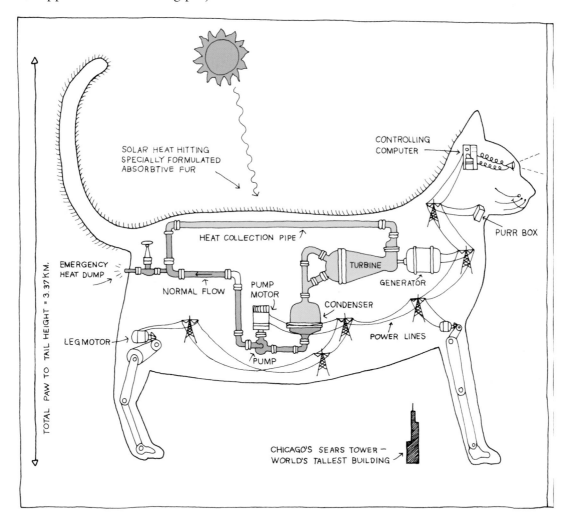

Figure 3.14 Originally Proposed 100 Megaton Solar Demonstration Solar Cat astride the (then) tallest building in the world.

3.5 THE SOLAR CHIMNEY

As the rest of the world surpasses the U.S. in its rate of solar and renewable energy development, they have also begun to vie for supremacy in the field of project grandiosity. A good example of this is the "Solar Chimney." It's primarily European proponents hope to build really big chimneys in deserts which would produce electricity by making an artificial wind of rising hot air. Wind turbines in the base of the chimney would extract energy from the moving air and turn it into electricity. Almost twice as tall as the tallest structure ever built, these colossal chimneys would dwarf all but America's partly completed 100 Megaton Solar Demonstration Cat.[16]

Figure 3.15 The proposed "Solar Chimney" showing how air is heated as it flows through the circular glass-roofed greenhouse structure around its base, then up and out the top of the chimney. A 100 Megaton Solar Demonstration Cat with other ideas is shown for scale.

16. The current world record holder for project grandiosity.

3.6 PHOTOVOLTAICS

Photovoltaic cells convert sunlight directly into electricity with only subatomic moving parts. Their inner workings can be explained quite simply by imagining that sunlight comes in very small, separate and discrete units called "pho-toms."[17] Each pho-tom which hits the surface of the cell completely vanishes upon impact, but leaves behind a tiny 'hole' into which nearby electrons fall. The movement of these electrons creates an electric current which is then channeled to an external circuit. So the word "photovoltaic" essentially means turning pho-toms into volts, or sunlight into electricity. Photovoltaic cells produce "direct" current which is changed into "alternating" current[18] for most uses.

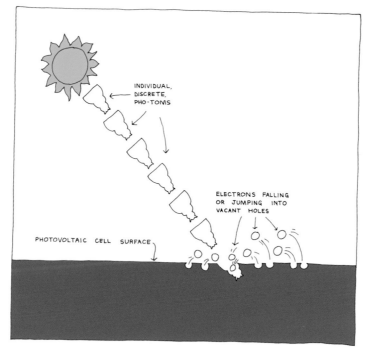

Figure 3.16 Movement of electrons into "holes" causes electric current in photovoltaic cells

17. Human scientists are still pondering over the exact nature of light, the concept of "pho-toms" being similar to what they call "photons."

18. Scientists are also pondering over the theoretical differences between "alternating cats" and "direct' cats," from which the possibility arises that science has in fact ceased to advance, since what could a cat possibly alternate into that was better than a cat?

Photovoltaic Cells, Modules, Arrays, and the Meaning of PV Photovoltaic cells
are pretty small, and do not make much power by themselves, but they can be hooked
together to form bigger units called modules.[19] Modules can in turn be hooked together to
form arrays, which are normally the major and most visible part of a photovoltaic system.

Some people do not like the word "photovoltaic." It just has too many syllables, and when
people first learn how simple and elegant this technology is, they are aghast at having to
pronounce such a mouthful. Even the people who work with these things every day usually
use the pronounced abbreviation of "PV." So they would say PV cells or PV modules, PV
arrays, or PV systems.[20] Nevertheless, newcomers to this technology almost always first use
the term "solar cells" not realizing that a solar cell could be something entirely different.

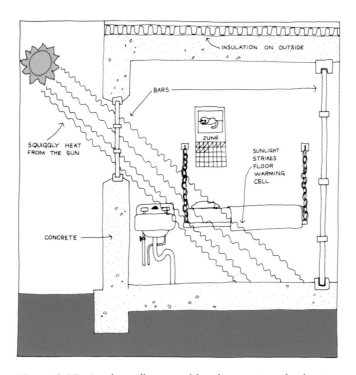

Figure 3.17 A solar cell warmed by direct gain solar heating
does not produce electricity.

19. Photovoltaic "modules" can be thought of as the basic building block of photovoltaic systems, and
    can be anywhere from a few centimeters to a few meters square.

20. If you don't like to say "photovoltaic" or "PV," use the term "solar-electric panels." In addition to
    being easier to pronounce and less "cliquish," it distinguishes solar-electric from solar-thermal
    (or solar heating) panels and systems.

No matter what you call it, photovoltaic technology is pretty fantastic. It is clean, quiet, works well in small to large sizes, doesn't weigh much or even look all that bad, and it works well almost everywhere. These admirable traits may be due to its similarity to the way cats chose as their preferred way to make electricity as described on the next page.

People first developed photovoltaics as a power source for things in very remote locations, like outer space, where batteries, generators, and fuel or transmission lines were just too hard to use. At first the cost of photovoltaics was very high. Then people found that if they used just little bits of it, the cost wasn't all that much. They began using it to power little things like calculators and wristwatches, then bigger things like ocean buoys, roadside call boxes, and mountaintop telecommunication stations. And as the cost has dropped, the range of cost-effective photovoltaic applications has grown ever more rapidly, causing photovoltaic proponents to become so exuberant they may be soon in danger of applying this technology in circumstances where something else would probably work better.

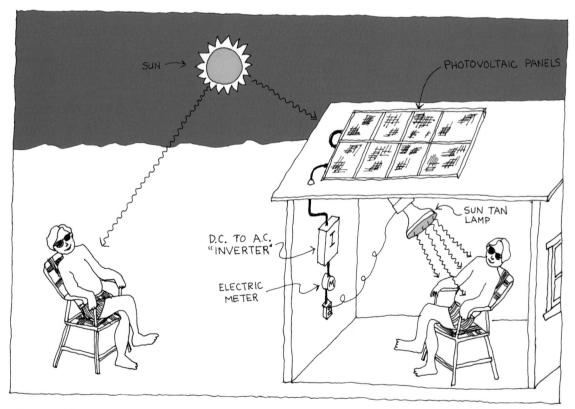

Figure 3.18 A circumstance where photovoltaic technology may not be the best choice.

3.7 DI-MEOWIUM OXIDE SOLAR ELECTRIC CATS

A di-meowium oxide[21] solar electric cat is quite simple in theory yet performs with unfailing reliability. The key to its successful operation is in the di-meowium oxide coating which is applied to the cat's fur. Meowium is a naturally occurring element which people have not yet discovered. Its location on the Periodic Table of the Elements is in a place where human scientists never thought to look. Almost all scientific effort has gone into finding new, heavier elements by throwing smaller elements together in atom smashers and hoping they will stick together. Meowium is found by looking in the other direction, toward things that are so light and fanciful it is a wonder they exist at all!

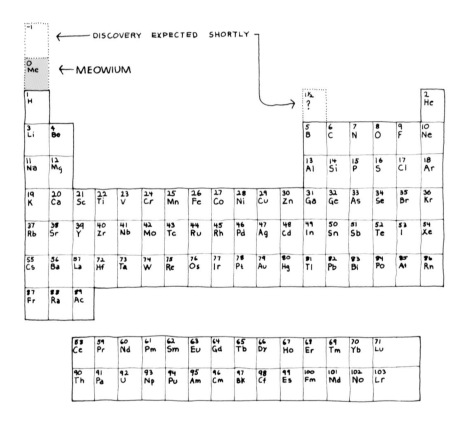

Figure 3.19 Periodic Table of the Elements showing new elements discovered or about to be discovered by cats. (Note how overall shape of table is moving toward it's ultimate catlike configuration.)

---

21. The first edition of this book incorrectly named this compound "meowium di-oxide" while expressing its chemical formula as "MeO$_2$." Thanks to a brilliant researcher at the California Institute of Technology, this error can now be corrected.

When combined with oxygen to form di-meowium oxide (Me₂O) it becomes a very durable coating exhibiting remarkable properties when applied to a cat's fur. When di-meowium oxide is applied to the fur of a cat with color variations in its coat (such as a tabby cat) an electric voltage is produced between the different colored areas in the presence of sunlight. Electrons move from dark to light fur areas and their movement causes an electric current. In a properly bred solar electric cat, this current can be channeled to charge special portable batteries. The stored power can then be periodically discharged by plugging a special tail adapter plug into any household outlet. The energy could then be injected into the electric utility's power lines for use elsewhere.

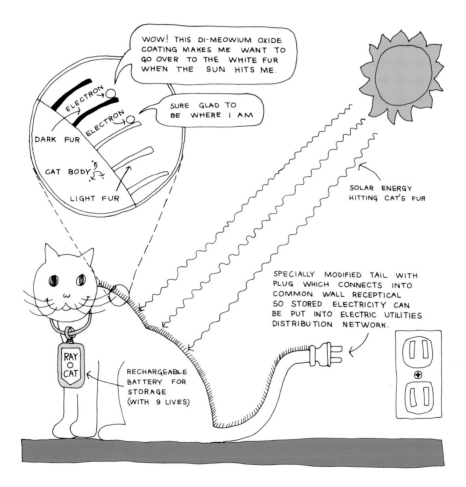

Figure 3.20 A Di-Meowium Oxide Solar Electric Cat with battery storage and tail-plug inverter for grid intertie.

Electricity produced by di-meowium oxide solar electric cats is amazingly inexpensive. This is because a little meowium goes a long way and is very cheap to begin with. In addition, solar electric cats of this type are generally 90 to 98 per cent efficient! Compare this to the most advanced human photovoltaic cells, still struggling to maintain 15 percent conversion of solar energy into electricity.

All this makes solar electric cats very attractive to people. Cats realize this, but will not allow people in on their central secret: meowium. Cats fear that people would stop at nothing to exploit this energy source. The mere idea of people raising huge herds of ill-fed outdoor cats mercilessly plugged into electric outlets[22] merely for production of electricity is a thought no responsible cat can allow to come to fruition.

Figure 3.21   Many unhappy di-meowium oxide solar electric cats in the desert plugged into the electric distribution network.

---

22. Or inlets, depending on your point of view.

3.8 HYDROGEN: THE MEOWIUM ALTERNATIVE?

Hydrogen has been mistakenly considered by some people to be a good substitute for meowium. Some even think of hydrogen as the human meowium! However close it may be to meowium on the Periodic Chart of the Elements, hydrogen is <u>not</u> meowium, and it never will be! Nevertheless, hydrogen is a fantastic fuel. When it burns, it makes just heat and water, and none of those pesky polluting by-products of fossil or even bio-fuels.

The only trouble with hydrogen is that it does not exist by itself in nature. You can't just dig a hole in the ground and expect hydrogen to squirt out.[23] Like meowium, it is very plentiful, but just about all hydrogen is locked up in chemical bonds with other elements, as in water, for instance, where it is combined with oxygen.

Figure 3.22 Person waiting for hydrogen to squirt out of the ground after digging a hole.

It takes energy to extract the hydrogen from water or anything else, and the only real question is where does that energy come from? Wind electric generators?[24] Parabolic Trough Solar Power Plants? Photovoltaic systems?[25] Biomass? Coal? Oil? Natural gas?, Nuclear power plants?[26]

---

23. This is not exactly true since you can dig a hole in the ground and expect hydrogen to squirt out. You will simply be disappointed when it doesn't!

24. This is a great application for wind electric generators, since locating hydrogen electrolysis plants near wind farm sites would eliminate the need for transmission lines to send electricity off to cities.

25. Parabolic Trough plants could do it, so could PV, but neither is quite as nifty a match as wind.

26. Sure, bio-fuels could do it, as could fossil fuels, or even nuclear, but there you go again with all those pesky polluting by-products!

4. SOLAR CAT ECONOMICS

The economics of solar cats are easy to understand. The renowned economist, Dr. Alan Catspan, best explains this in the semi-annual, 12-word speech he periodically gives to the International Solar Cat Society.[1]

Human solar economics are just about impossible to understand. They are just too complicated. This does not mean that human solar technology is bad or too hard to understand. It simply means that people have a hard time figuring out what <u>is</u> good,[2] and how to value it when they do see it!

Figure 4.1 Dr. Alan Catspan, chairman of the International Solar Cat Society's Feral Reserve Commission, giving his famous 12 word speech explaining the economics of solar cats.

1. This speech is given once each year to cats in the Southern Hemisphere on December 21, and once each year to cats in the Northern Hemisphere on June 21.
2. The ability to recognize what is good and what isn't is not only fundamental to economics, but is also fundamental to fundamentalism itself!

4.1 THE VALUE OF SOLAR ENERGY

Solar energy is valuable. The first step toward recognizing how valuable it actually is—especially compared to other things—is to know the difference between energy and power. Energy is *how much* work can be done. Power is *how fast* that work can be done.

Figure 4.2 A single government employee can lift a one-pound stack of forms from a desk top to a shelf one foot above desk level in six hours. If 21,600 identically capable civil servants could somehow work in unison (with no loss of efficiency) to accomplish the same task, only one second would be required. The same amount of energy was expended in each case, however the multitude of paper lifters applied much more power to the same task.

The Metric Cat Power is often measured in kilowatts,[3] and electric energy in kilowatt-hours. To appreciate how much power is in sunlight, the consider the concept of a "metric cat." About one kilowatt of solar power falls on a metric cat whose one square meter absorbing area is oriented perpendicular to the sun's rays in bright sunlight. In one hour, one "kilowatt-hour" of solar energy would hit that cat. In most places, a rate of one kilowatt per square meter occurs near the middle of the day. The rate at which solar energy hits anything is less in the morning and afternoon, and is much much less at night.[4]

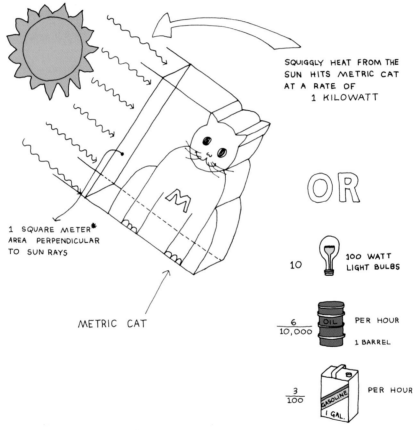

Figure 4.3 The Metric Cat.

---

3. There are still a lot of people who use the British measure of the "Horsepower." A kilowatt is not quite as as big as a horsepower, possibly due to 25% bigger horses in Britain compared to those in France where the metric system of measurement was first developed.

4. Weather also has a lot to do with how much sun can hit a cat.

How Much Solar Energy Is There? There is a lot of solar energy.[5] About 1,600 kilo-watt-hours of solar energy fall on a typical, horizontal square meter of absolutely average America each year. If this much energy was converted entirely into gasoline, (as with horizontal metric gas cats) enough gas would be produced to fuel a 30 mpg car 1,360 miles. An average American car owner would merely need several such cats in order to stay happily rolling along. Of course, no one lives in an absolutely average American place, nor has anyone yet perfected a 100% efficient metric gas cat! Nonetheless even people in such cloudy places as London or Berlin would need only about 1.5 times as many metric gas cats to make the same amount of gas.

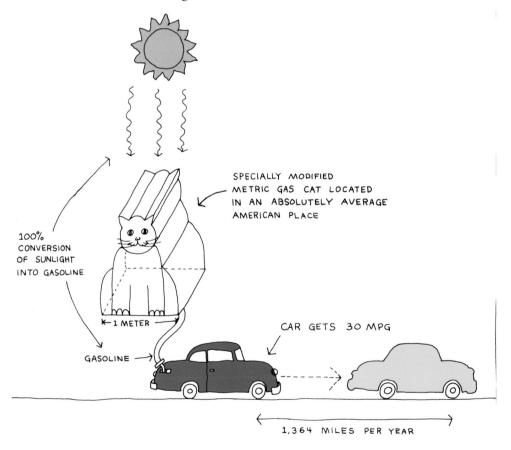

SPECIALLY MODIFIED
METRIC GAS CAT LOCATED
IN AN ABSOLUTELY AVERAGE
AMERICAN PLACE

100%
CONVERSION
OF SUNLIGHT
INTO GASOLINE

←1 METER→

GASOLINE →

CAR GETS 30 MPG

1,364 MILES PER YEAR

Figure 4.4 Filling car gas tanks with gasoline made by specially modified metric gas cats.

5. See chapter 1, footnote 5.

Measuring Sunlight Cats have no need to measure the strength of sunlight. They are born knowing this kind of thing. But as people begin to study solar energy, many become curious about how much of the sun's energy actually hits the ground where they live and hits the things they might make to collect it. People have devised different ways to measure sunlight, none of which is both highly accurate and economical.[6] The most accurate way to measure the strength of sunlight is with a purranometer.[7]

Figure 4.5 A purranometer consisting of a cat in a wind-sheltered, transparent dome absorbing solar energy and purring at a frequency proportional to the strength and goodness of the sunlight that strikes it. The cost of the human observer counting purrs can be minimized if rich volunteers who don't need to be paid are employed.

6. The most economical way to do this is with satellites. However, such estimates can never be as accurate as instruments located right on the ground where solar energy actually does its hitting.

7. The name "purranometer" should not be confused with "pyranometer" which is what people call the little devices they make to measure the strength of sunlight.

4.2 SOLAR THERMAL CAT ECONOMICS

As discussed earlier, two kinds of solar thermal cats are of practical interest: "passive" cats and "active" cats. With both kinds, initial purchase and installation costs are very small. Operating and maintenance costs are modest for each kind, with passive cats costing less for veterinary care and food.

Peoples' solar thermal systems are not nearly as economical as solar cats, but then, what did you expect? In comparison, peoples' systems usually have high first costs, low operating and maintenance costs, and produce modest amounts of energy. As with cats, peoples' passive solar thermal systems do somewhat better than the active kind.

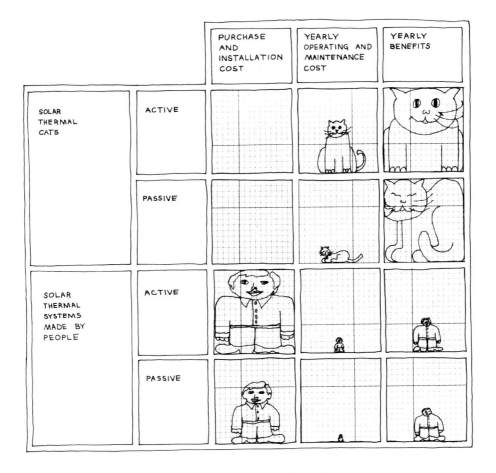

Figure 4.6 Comparing the costs and benefits of cat's and peoples' solar thermal systems.

4.3 SOLAR ELECTRIC CAT ECONOMICS

Of peoples' methods for solar electric conversion, only hydroelectric power has been seriously developed on a large scale. Other methods described in Chapter Three are really just beginning the race toward market development. Though di-meowium oxide solar electric cats won this race before it even started, it will be interesting to see who comes in second.

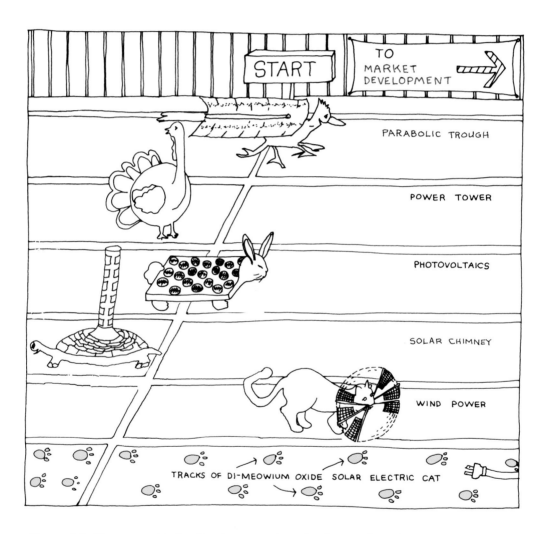

Figure 4.7 The race to Market Development where the distance from the starting line to the leading nosetip of the technaminal represents the total amount of electric energy generated worldwide since the beginning of time by that technaminal.

The Promise of Solar Technology People have long hoped and expected that scientific breakthroughs will lower the cost of solar energy technologies so much that everyone will start using them. Then the world will become free of dependence on oil and other polluting technologies, and everything will be wonderful. Great cost reductions have occurred—especially with photovoltaics where costs are less than 10% of what they were two decades ago.[8] While further reductions are likely, they will continue to be driven as much by government policies as by pure "market forces."

Figure 4.8 People witnessing a breakthrough without knowing it.

8. Taking inflation into account.

4.4 THE TOOLS OF ECONOMICS

Income Cats Deductions, Credits and Cuts[9] Some governments use tax laws or even rebates to increase solar energy use. When a solar system is installed, income taxes are reduced or rebates are paid to the buyer. This reduces the first cost of solar systems making their payback times shorter. A similar credit on veterinary bills and cat food costs has been proposed for each cat that comes into a home. Such an "income cats credit" would pay for all operating and maintenance costs of the cat in question, as long as it was counted. One drawback of this proposal is the need to form yet another government bureaucracy in order to count cats.

Figure 4.9 Employee of the Federal Income Cats Counting Service at work.

9. Rebates are easy, credits are ok, as are deductions, but Cats Cuts take the cake! Who in their right mind could ever even dream that either reducing the number of cats that enter one's life or cutting off undesirable parts of those cats who do enter could ever possibly come to good?

Payback Time Since the benefits of solar cats so greatly outweigh the costs, they are obviously good investments. Peoples' solar systems take longer before their benefits surpass their costs. The time it takes for this to happen is called the "Payback time" and is easy to figure out as shown below.

Figure 4.10 The time to recover or "pay back" the cost of initial investment in a human-devised solar energy system is initial cost to install it divided by the yearly energy cost savings minus the yearly operating and repair costs.

Inflation Many years ago, economists invented inflation as a way to achieve job security. Inflation makes everything more complicated, and causes a great demand for ever more professional economists. Inflation, however, makes solar thermal systems look even more attractive, because the benefit of solar system operation (fuel cost savings) inflates faster than everything else! Solar tcats (and actually peoples' systems as well) are therefore an excellent hedge against inflation.

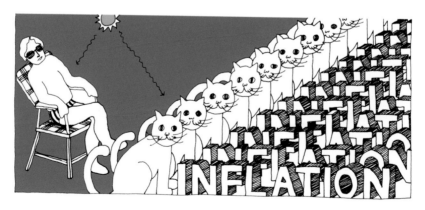

Figure 4.11 Solar Cats as a hedge against inflation.

5. CHOOSING A SOLAR CAT

There are many ways to choose a solar cat. However picking a good one is not easy as countless variables complicate a clear choice. The same problem arises when people try to choose solar electric or solar heating systems. In either case, it is best to look at each part of the cat or system and, using common sense, see if it seems adequate in view of what it is supposed to do. If common sense doesn't work, there are always experts to help out. Experts on any subject are always nearby, but choosing an expert is equally difficult, and should certainly not be left to blind chance.

Figure 5.1 Person choosing a solar cat.

Figure 5.2 Person choosing an expert.

The Automotive Technique   There are many good books which tell about things to consider when selecting a human-designed solar system. A few of these books are listed in the Appendix. In addition to reading, a good way to find out about a particular system is to talk to someone who has owned and operated one for a few years. In this way, the selection process becomes very similar to that of buying a new car.[1]

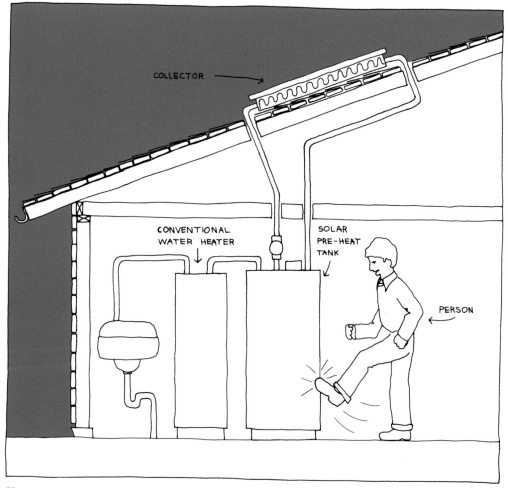

Figure 5.3 Person using an automotive evaluation technique on a solar water heater.

---

1. Similar and equally effective application of this same technique can apply to evaluating solar electric systems by kicking inverters, turbine generators, and the bases of solar chimneys or wind turbine supports.

The 10-Day Free Home Trial Another method for evaluating and choosing a solar cat is the 10 day free home trial. With this method, a kitten is brought home for a 10 day period during which the prospective owners get a chance to examine it in the privacy of their own home. This method also works well for kittens in search of a home as few people are able to resist falling into "kitty love" after about ten minutes. An interesting aspect of this selection method is that it really forces the prospective owners to examine their reasons for wanting a cat in the first place.

Figure 5.4 People trying the 10-day Free Home Trial.

The Solar Cat Evaluation Checklist Many people do not have the time to read books, to choose (or listen to) experts, to embark on 10 day free home trials, or to think for themselves. The Solar Cat Evaluation Form is made especially for these people. It is particularly useful in comparing two or more cats by setting absolute standards for goodness from "very bad" to "very very good." To use the checklist, carefully determine the correct rating for each evaluation item for the cat under consideration. Then determine the Solar Cat Evaluation Number and Evaluation Rating Classification.

SOLAR CAT EVALUATION FORM

To determine the Solar Cat Evaluation Number (SCEN), first enter the correct number from 0 to 10 for each of the rating items (A. through O.), insert the rating values into the following equation and add, multiply or divide as appropriate. Ratings are 1 to 10 or as indicated.

$$\text{SCEN} = A \times E \times F \times G \times O \times \left(B + C + D + H + I + J + K + L + \frac{1}{(M+1)} + N \right)$$

A.	Is it really a cat? (10 is very much a cat)	
B.	Number of paws (insert actual number of paws)	
C.	Width of left front paw (in inches)	
D.	Length of tail (in paws)	
E.	Maximum cross sectional area (in square paws)	
F.	Weight of cat (undressed and dry in pounds)	
G.	Fur color (black = 10, white = 1, transparent = 0)	
H.	Number of stripes (if no stripes, enter 0)	
I.	Width of stripes (in paws, if no stripes, enter 0)	
J.	Number of fur colors	
K.	Maximum purr rate (in purrs per minute)	
L.	Average fur length (in paws)	
M.	Fear of dogs (10 is very afraid)	
N.	How much it likes to eat (10 is a lot)	
O.	How much it likes to lie in the sun (10 is a lot)	

There are several ways to interpret the result of this equation. One way is to consider it nonsense and get the cat you like the most. This method is considered quite reliable. But for those who don't know what they like, it is generally considered that high values of "SCEN" are best, and low or negative values are not so good. If you come up with imaginary numbers (and there really are such things), you have great potential as a solar cat owner, but not as a mathematician. It is certainly easy to compare the SCENs of two cats under review, the cat with the higher SCEN being the better bet. If a great many cats are to be evaluated, the "SOLCAT" computer program may be of value. The SOLCAT program is explained, and a program listing is presented in Appendix 2.[2] If only one cat is under review, its SCEN can be compared to the Standard Solar Cat Evaluation Rating Classifications as presented in the following table:

Table of Standard
SOLAR CAT EVALUATION RATING CLASSIFICATIONS

SCEN	Rating Classification
0 and below	Very bad
Above 0, up to 10,000	Not so bad, not so good
Above 10,000 up to 1,000,000	Fair
Above 1,000,000 up to 10,000,000	Pretty good
Above 10,000,000 up to 100,000,000	Very good
Above 100,000,000 up to 1,000,000,000	Very, very good
Above 1,000,000,000	Unbelievable! (but not unusual)

2. An interactive version of this program is also available on the Internet for those who either can't read, can't think. can't use pencils, can't add, can't multiply, or just can't seem to anything without doing it on-line. It is at www.solarcat.com

Cats are self-sustaining. They are well known for their ability to take care of themselves. They are also the only known purrpetual[1] life form. People are not quite as sustainable due to their difficulty with long term thinking.[2]

Even though cats don't really need any help sustaining themselves, it is a good idea for people to help cats sustain themselves.[3] The remainder of this chapter presents information about how people can most effectively operate, train, and maintain a solar cat, as if they actually had a vital role in this regard.

Figure 6.1 Person learning how to sustain a solar cat by reading *The Return of The Solar Cat Book* to a companion who is simultaneously helping to sustain the person.

1. This refers to the phenomenon where a cat purrs when a person pets it.
2. People do recognize the value of long term thinking, but usually forget whatever long term thoughts they have shortly after thinking them.
3. This is a good idea for two reasons: first it would keep people from doing other things that were not nearly as good, and second, they might learn something about how to be as sustainable as cats are.

Operation Cats are fully automatic and need little or no input from anyone in order to function efficiently. However, they can break down so it is advisable to become familiar with how cats work so that substandard performance is quickly identified and corrected. Learning about cats can be accomplished by reading books about cats, and by going to classes, seminars and lectures about cats. Direct observation of cats "in the field" is also of great value, there being no academic substitute for practical experience.

Substandard solar cat performance can occur either because the cat forgot what it was supposed to do, or it got sick, hurt, or otherwise damaged. If the problem can be traced to cognitive dysfunction, a cat training program possibly including medicinal catnip may be in order. If the cat is sick, hurt, or damaged, the services of a professional cat repair specialist may be in order.

Figure 6.2 Engaging the services of a professional cat repair specialist.

Training Contrary to popular belief, it is possible to train a cat. The greatest success has always been in training them to do things they were about to do anyway. However, limited success has been reported in other areas. Cats learn some things very quickly, such as how to purr, where and when cat food appears, or how to proceed (or recede) as menacing dogs approach. Other things are much more difficult for them to pick up, such as how to drive trucks or become successful prizefighters.

Figure 6.3 Cat being trained how to purr (successfully).

Figure 6.4 Man in cat suit jumping rope trying (unsuccessfully) to get cat to do the same.

If a cat does not like to sit in the sun to absorb energy, it is sometimes possible to bring the sun to the cat, either with mirrors, or clever cat food dish positioning.[4] If such devious methods do not work, more effort should be devoted to proper cat selection.

Figure 6.5 Cat having more important things to do than to sit in the sun.

Figure 6.6 Cat absorbing sunlight while eating.

---

4. Never keep cat food in the sun. as it will spoil quickly. It is the cat you want to sit in the sun, not the cat food!

Maintenance   Proper maintenance of a solar cat is best accomplished with strict adherence to a properly formulated maintenance schedule. Since each cat has different servicing needs, no single maintenance schedule applies in all cases. The sample schedule presented here should therefore be reviewed and modified as necessary by a professional cat maintenance specialist, in view of the cat under consideration.

(Sample) Solar Cat Maintenance Schedule

Service Interval	Item	Remove	Suggest	Love	Check level	Clean	Refill	Inspect	Service
12 hours or 12,000 purrs [1] *(whichever occurs first)*	cat			X					
	water dish				X		2		
	food dish					X	X	X	
24 hours or 24,000 purrs *(whichever occurs first)*	cat			X				3	
	water dish					X	X		
	food dish					X	X		
	litter box					X		4	
	lizards in bed	X						X	
	bird feathers	X						X	
	fur					X		X	
Every second litter box change	go outside [5]		X						
	paw rotation [6]								X
7 days or 168,000 purrs *(whichever occurs first)*	ears							X	2
	teeth							X	2
	fleas	X							
1 month or 720,000 purrs *(whichever occurs first)*	cat			X		X		X	X
1 year or 8.76 megapurrs *(whichever occurs first)*	cat							7	7

1. A purrometer can be used to count purrs.
2. If necessary.
3. General external inspection for appearance, attitude, and appetite.
4. Inspect cat's stool (and urine if possible) for abnormalities.
5. Applies to cats who are allowed outside.
6. Applies to cats who have removable and interchangeable paws.
7. Inspection and servicing by a cat maintenance professional.

APPENDICES

Appendix 1. ACTIVE SOLAR THERMAL CAT ECONOMIC CALCULATIONS

The following calculations[1] apply to an active solar thermal cat system as shown above.

1. A Standard Cat[2] whose body temperature is 100.5°F goes outside to warm up in the sun[3].

2. The sun hits each square foot of the cat's fur at a rate of 300 BTU per hour.

3. Standard Cat fur is especially absorptive and collects 90% of the solar energy that hits it. This results in: 300 x 0.90 = 270 BTU absorbed per hour per square foot of cat fur

4. The area of cat fur exposed to sunlight in a plane perpendicular to the rays of the sun is 1.33 square feet, so that: 270.0 (BTU per hour is absorbed per square foot of cat) x 1.33 (square feet of fur per cat) = 359.1 BTU per hour absorbed per cat.

5. The cat absorbs solar energy until its body temperature reaches 102°F, the cat then goes indoors to discharge the collected energy to the interior of the house. Since it starts at 100.5°F, its temperature has increased 1.5°F, and in so doing it absorbs 15 BTUs. At a rate of 359.1 BTU per hour it will take 2.51 minutes to heat the cat from 100.5° to 102.0°F, since: 15 (BTU per cat) x 60 (min. per hr.) / 359.1 BTU per hr. per cat = 2.51 minutes

6. If the cat takes 3 minutes for its outdoor trip, and 7 minutes to discharge the heat to interior of the house, 10 minutes are required to complete each round trip cycle.

7. When inside, the cat discharges both the stored solar heat collected in its body of 15 BTU, and the catabolic heat generated by normal cat body functioning. The catabolic heat generation rate can be estimated from the equation presented in the section on catabolism in Chapter Two which predicts a 92.8 BTU per hour catabolic heat generation rate for the standard cat. Since the cat spends 7 minutes indoors discharging heat, the total catabolic heat put into the house is: (7 min.) x (92.8 BTU per hr) / (60 min. per hr.) = 10.8 BTU. When the solar heat input of 15 BTU is added: 10.8 + 15 = 25.8 BTU is put into the house during each complete charge and discharge cycle.

---

1. All costs are expressed in 1979 U.S. dollars. (Hint: 2002 dollars are 4 times as worthless.)

2. See Appendix Part 3..

3. The normal body temperature cat can vary from 100.5°F to 102.0°F throughout a day.

8. Since each cycle is 10 minutes long, 6 can be completed each hour. If 8 hours of cycling can occur each day of a 250 day heating season, the yearly heat input to the house as a result of this cycling is as follows: 6 cycles/hr. x 8 hours/day = 48 cycles/day 48 cycles/day x 250 days/yr. = 12000 cycles/yr. 12,000 cycles/yr. x 25.8 BTU/cycle = 309,600 BTU/yr.

9. Since the cat is on its "solar cycle" only 8 hours per day, the remaining 16 hours per day can be spent hanging around inside the house discharging catabolic heat at a rate of 92.8 BTUs per hour, so that: 92.8 BTU/hour x 16 hours/day = 1484.8 BTU/day is input to the house by night time catabolic heat generation. In a 250 day season, this comes to: 1484.8 BTU/day x 250 days/year = 371,200 BTU/year.

10. If we add the daytime and nighttime heat inputs: 309,600 BTU per year (Daytime) plus 371,200 BTU per year (nighttime) = 680,800 BTU per year supplied by the solar cat.

11. If this energy was supplied instead by electric resistance heating, at a cost of $0.10 per kilowatt hour, the value of the solar cat energy would be $19.95 per year. While $19.95 per year per cat is nothing to sneeze at, normal operating and maintenance costs certainly are! These include food, flea collar, and veterinary costs as follows:

    A. Flea collars: $3.98 per collar x 3 collars per year = $11.94 per year per cat

    B. Food: 0.5 cans of wet food per day per cat at $0.30 per can yields $0.15 per day per cat plus 50 grams of dry food per day per cat at $0.0012 per gram yields $0.06 per day per cat. So, total yearly cat food cost is: $0.15 + $0.06 = $0.21 x 365 = $76.65

    C. Veterinary Care: These costs vary from zero to several hundred dollars per year. Therefore, we shall use $110.91 per year per cat for this cost. Total operating and maintenance costs are then: $11.94 + $76.65 + $110.91 = $199.50

12. Operating and maintenance costs are therefore ten time the yearly savings! This obviously will not do if solar cats are to be really feasible! Fortunately, several things can be done to make the picture look a whole lot better:

    A. Use "passive" cats to reduce veterinary and food costs. Passive cats also add more catabolic heat since they seldom go outside.

    B. Consider the additional energy input due to "Latent Purrs" as explained in Chapter Two. This item alone could increase the output of a sufficiently happy solar cat by a factor of 100 or more!

    C. Reinstate the federal income cats credit, give out rebates, negative percent financing, cash-back, income cats cuts, eliminate the breath[4] tax, thereby eliminating all direct costs to the cat owner, while spreading them out over all of society and to future generations!

    D. Let them eat cake! This famous European solution would effectively eliminate food costs, if not reality as well!

---

4. This is a term recently coined by political linguistics to refer to the Cat's Breath Tax where the love and perseverance of loyal humans is severely taxed with each revoltingly stinky, fishy smelling breath a cat exhales, into the face of its human companion.

Appendix 2. "SOLCAT" SOLAR CAT EVALUATION COMPUTER PROGRAM

The SOLCAT computer program goes through the same simple checklist and evaluation procedure which appears in Chapter 5. Though most 6-year-old humans (with a pencil or crayon) could easily use the "long hand" written checklist method, the SOLCAT program allows use of the latest in sophisticated electronic equipment to accomplish the same task! The SOLCAT computer program was first written on a Data General model S-130 minicomputer in a version of the BASIC computer language. Other computers may have slightly different versions of the BASIC language, so the program listing as presented here might not work properly at first try.

SOLCAT Program Listing

```
0005 DIM V$(40), Z$(25)
0006 PRINT
0007 PRINT
0010 PRINT "Welcome to SOLCAT-Version 1.0"
0020 PRINT
0030 PRINT "This program aids in selecting a solar cat."
0040 PRINT "Answer the following questions about the cat"
0050 PRINT "under evaluation.  After answering all questions"
0060 PRINT "the program will automatically tell you how good"
0070 PRINT "or bad the cat is.  It then allows you to start"
0080 PRINT "all over again in case you want to get a"
0090 PRINT "different answer, or have another cat to"
0100 PRINT "evaluate."
0105 PRINT
0110 PRINT
0120 PRINT "All questions should be answered with a number"
0130 PRINT "from 1 to 10 unless otherwise indicated."
0140 PRINT
0150 PRINT
1000 INPUT "Name of cat (spell it out)                        =", V$
1001 PRINT
1002 INPUT "Is it really a cat? (10 is very much a cat)       =", A
1010 INPUT "Number of paws                                    =", B
1020 INPUT "Width of left front paw (in inches)               =", C
1030 INPUT "Length of tail (in paws)                          =", D
1040 INPUT "Maximum cross sectional area (in square paws)     =", E
1050 INPUT "Weight of cat (in pounds, undressed and dry)      =", F
1060 INPUT "Fur color (black = 10, white = 1, transparent = 0)=", G
1070 INPUT "Number of stripes (if no stripes, use 1)          =", H
1080 INPUT "Width of stripes (in paws)                        =", I
1090 INPUT "Number of fur colors                              =", J
1100 INPUT "Maximum purr rate (in purrs per minute)           =", K
1110 INPUT "Average fur length (in paws)                      =", L
1120 INPUT "Fear of dogs                                      =", M
1130 INPUT "How much it likes to eat                          =", N
1140 INPUT "How much it likes to lie in the sun               =", O
1150 Q7 = 0
1200 PRINT
```

```
1220 PRINT
2010 S1 = A* E* F* G* O* (B+C+D+H+I+J+K+L+(1/(1+M))+N)
3000 IF S1 = 0 THEN Z$ = "VERY BAD"
3010 IF S1 < 0 THEN Z$ = "VERY BAD"
3020 IF S1 > 0 THEN GOTO 5000
3100 PRINT USING "SOLAR CAT EVALUATION NO = ###,###,###,###,###,###.##", S1
3110 PRINT
3120 PRINT "SOLAR CAT STANDARD RATING =", Z$
3124 IF Q7 = 0 THEN GOTO 3130
3125 PRINT
3126 PRINT "Unbelievable! (but not unusual)"
3127 PRINT
3128 PRINT "Nevertheless, GET IT QUICK"
3130 PRINT
3150 INPUT "Do you want to start over again? (Y or N)", Y$
3155 PRINT
3156 PRINT
3157 PRINT
3160 IF Y$ = "Y" THEN GOTO 1000
3170 IF Y$ = "y" THEN GOTO 1000
3180 IF Y$ = "YES" THEN GOTO 1000
3190 IF Y$ = "yes" THEN GOTO 1000
3200 IF Y$ = "Yes" THEN GOTO 1000
3210 GOTO 9999
5000 IF S1 = 10000 THEN GOTO 6000
5010 IF S1 < 10000 THEN GOTO 6000
5020 IF S1 = 1000000 THEN GOTO 6100
5030 IF S1 < 1000000 THEN GOTO 6100
5040 IF S1 = 10000000 THEN GOTO 6200
5050 IF S1 < 10000000 THEN GOTO 6200
5060 IF S1 = 100000000 THEN GOTO 6300
5070 IF S1 < 100000000 THEN GOTO 6300
5080 IF S1 = 1000000000 THEN GOTO 6400
5085 IF S1 < 1000000000 THEN GOTO 6400
5090 GOTO 6500
6000 Z$ = "NOT SO BAD, NOT SO GOOD"
6010 GOTO 3100
6100 Z$ = "FAIR"
6110 GOTO 3100
6200 Z$ = "PRETTY GOOD"
6210 GOTO 3100
6300 Z$ = "VERY GOOD"
6310 GOTO 3100
6400 Z$ = "VERY VERY GOOD"
6410 GOTO 3100
6500 Q7=1
6520 GOTO 3100
9999 END
```

Appendix 3. CAT THERMAL UNITS (CTUs) — AND HEAT

Like the "BTU" or British Thermal Unit, the CTU is a measure of heat energy[1]. But the CTU is a more than just energy! Both the CTU and BTU[2] are precisely defined amounts of heat. A BTU is enough heat to raise the temperature of one pound of water one degree fahrenheit[3].

There are two kinds of CTUs: small CTUs and big CTUs. A small CTU is exactly enough heat to raise the temperature of the standard cat one degree fahrenheit. The standard cat weighs ten pounds, is gray, has short hair, and is very good. Since cats are comprised primarily of cat food and water, one small CTU is about the same as 10 BTUs. A big CTU is the amount of heat produced as a result of normal body catabolism of the standard cat over all of its nine lives, assuming that the cat is always at rest, and lives 15 years in each life.

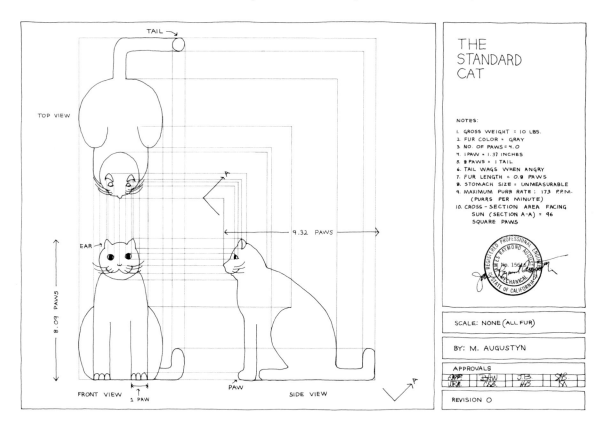

---

1. Cats invented the CTU long before anyone invented the British.

2. You can also call a whole bunch of BTUs a "Quad" if you have one quadrillion of them. "Big" energy thinkers sometimes use quads to express their big thoughts. A quad is enough energy to raise the temperature of the entire the Mississippi River 10°F.

3. There are several kinds of BTUs, the International Steam Table (IST) BTU is 4.1868 IST calories, and the Mean BTU is 4.19002 mean calories. CTUs are not mean, so the IST conversion applies.

Catabolism is discussed elsewhere, and is the part of body metabolism in which heat is released. There are 10,980,200 little CTUs in a big CTU. One big CTU also equals about 19.6 barrels of crude oil, yet CTU's are far from crude!

The key to understanding CTUs is that they are very anti-social. They do not like each other at all, and whenever they can, they try to go anyplace where there are fewer CTUs. You can always tell how many CTUs are in something by measuring its temperature. The higher its temperature, the more CTUs are inside the material in question. The sun is so grossly overpopulated with CTUs, that in order to get away, solar CTUs literally turn themselves into electromagnetic radiation and zoom out from the surface of the sun at the speed of light in all directions. Even if they end up in such a relatively cool place as on a cat's fur in what is left of the Amazon jungle, they are not satisfied. They still keep trying to get away from the nearby, though less numerous CTU population. CTUs are not completely happy until they reach the lonely depths of outer space where they can float alone without another CTU within millions of miles.

It is important to note that it is quite easy to hold onto CTUs if there aren't many of them around. The more CTUs there are, the harder it is to keep them in one place, as their distaste for each other grows with their numbers. This explains why it is so much easier to keep a house cold on a cold day than it is to keep it warm. When cats or people think about how to keep warm or cool, they are really thinking about how to get and keep the right number of CTUs nearby. Thus, the entire cat and human fields of thermodynamics and heat transfer are devoted to describing precisely how anti-social CTUs (and BTUs) are.

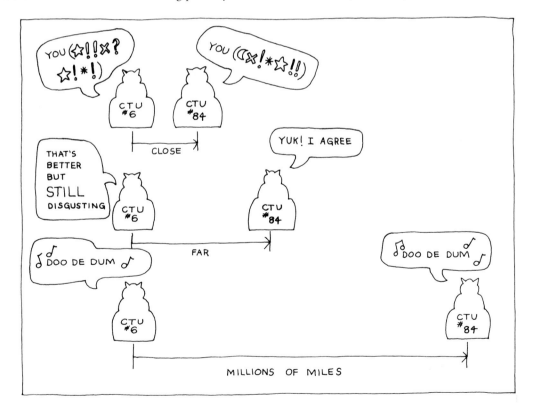

Appendix 4. CONCRETE CATS—THE ALLERGIC SOLUTION

Many people are allergic to cats. To such people, the thought of dozens of furry little electric or thermal cats running all over the house is disquieting at best. Solid concrete cats offer a remarkably good solution to this problem, as few people are allergic to concrete. Concrete cats can be painted dark colors and placed on south facing window sills to warm up during the day, then carried off to bed to keep sleepers warm at night.

Appendix 5. GLOSSARY

Advocat	A cat mathematician specializing in addition.
Bad	A cat who is not good.
Cat	The next best thing to the best thing there is.
Catabolism	The part of metabolism in which heat is released: "destructive" metabolism.
Catalyst	A cat analyst.
Catacomb	A special cat door which arranges cat fur in a desirable manner.
Cat Scan	A very penetrating and revealing look.
Cats' Cut	What cats get out of the deal (especially fat cats). (Also see Tax Cut.)
Cat Ion	When a cat stares at something.
Cata Tonic	Something good to drink.
Conservation	Virtue.
Catpacity Factor	The fraction of time a cat is awake and doing useful stuff.
Catastrophe	A silly human practice where a cat is kept as a trophy to impress other humans.
Ecomeowmix	Similar to the human field of Economics, but instead taking into consideration the <u>entire</u> mixture of meanings of the word "meow" and its subtle variations in evaluating something.
Efficiency	An inoffensive substitute for virtue. (See Conservation.)
Energy	Everything that isn't mass.
Electricity	A poor substitute for solar energy. Also a subtle vibration of the lumeniferous ether.
Gasoline	See Sex.
GigaPurr	A really really big purr.
Glop	In liquid form, an anti-corrosive additive to be mixed with potentially corrosive fluids. (For solid form see Yuk.)
Good	Most cats and many people.
Inflation	An imaginary notion in human economics in which peoples' energy and work becomes worth less as time goes by.
KiloPurr	A pretty big purr.
Kitty	A young cat.
Love	An essential ingredient in any successful solar cat.
Kitty Love	Like puppy love, yet softer and more serious.
Mega Purr	A really big purr.

Meow	Short for just about anything.
MicroPurr	A really small purr.
MilliPurr	A pretty small purr.
Money	Time, also energy as well as everything that isn't energy.
NanoPurr	A really really small purr.
Nuclear waste	A source of really really long term job security.
Oil	See Gasoline.
Photovoltaics	A messy name for a neat technology.
PV	Abbreviation for photovoltaics.
Politician	A teflon-based life form.
Pretty	See Really.
Purr	A form of energy expressed by cats having a magnitude defined by depth and frequency, and a value beyond compare.
PPM	Abbreviation for "purrs per minute."
Purrometer	A device used to count purrs consisting of an other wise unemployed person with a stopwatch.
Purranometer	A device consisting of a cat purring at a rate directly proportional to the strength of the sunlight striking it. The most accurate purranometers are called absolute catity purrnometers and cost a lot of money.
Purraheliometer	A purranometer running on helium.
Purrpetual	A term referring to the unique ability of solar cats to sustain themselves by getting humans to pet them, thereby inducing purrs while accidentally sounding a little like the human word "perpetual."
Really	A thousand times whatever it is.
Renewable energy	A physical impossibility according to the first law of thermodynamics. (See Sustainability.)
Sex	The opiate of the masses. (See Gasoline.)
Sustainability	A metaphysical impossibility but a nice sentiment nonetheless. (See Renewable Energy.)
Sun	That which turns night into day.
Tax Cut	All purpose cure for any ailment; particularly those associated with raising money for political campaigns. (Also see Cats' Cut.)
Time	Money.
War	The most efficient waste of time and energy yet devised by humans.
Yuk	Solid glop. (See Glop.)

Appendix 6. RECOMMENDED READING MATERIAL

Behling, Sophia and Stephan. *Solar Power, the Evolution of Sustainable Architecture*. New York: Prestel Publishing U.S.A., 2000.

Brown, G. Z. and Mark DeKay. *Sun, Wind and Light: Architectural Design Strategies*, 2nd Edition. Hoboken, New Jersey: Wiley, 2000.

Daniels, Farrington. *Direct Use of the Sun's Energy*. New York: Ballantine Books, 1964.

Eisenberg, Bainbridge, and Steen. *The Straw Bale House*. White River Junction, Vermont, Chelsea Green Publishing, 1994.

Gipe, Paul. *Wind Energy Basics*. White River Junction, Vermont, Chelsea Green Publishing, 1999.

Guzowski, Mary. *Daylighting for Sustainable Design*. New York: McGraw-Hill Professional, 1999.

Perlin, John. *From Space to Earth—The Story of Solar Electricity*, Cambridge: Harvard University Press, 2002.

Magazine. *Renewable Energy World*, James & James (Science Publishers) Ltd., 35-37 William Road, London, NW13ER, United Kingdom)

Magazine: *Solar Today*, American Solar Energy Society, 2400 Central Avenue, Boulder, Colorado 80301.

Web site: American Solar Energy Society (ASES) http://www.ases.org

Web site: International Solar Cat Society (ISCS) http://www.solarcat.com

Web site: International Solar Energy Society (ISES) http://www.ises.org

Web site: U.S. Department of Energy, National Renewable Energy Laboratory http://www.nrel.gov